BAJA'S WILD SIDE
A Photographic Journey Through Baja California's Pacific Coast Region

Hope you enjoy these images, inspired by our Scripps shark research program in Baja!

A milestone of conservation photography, highlighting all that is at stake in the fight to save Baja's wild places.
—SERGE DEDINA, Executive Director of WILDCOAST and author of *Surfing The Border*

Baja's Wild Side

A PHOTOGRAPHIC JOURNEY THROUGH BAJA CALIFORNIA'S PACIFIC COAST REGION

DANIEL CARTAMIL, PhD

Sunbelt Publications, Inc.
SAN DIEGO, CALIFORNIA

Baja's Wild Side: A Photographic Journey Through Baja California's Pacific Coast Region

Sunbelt Publications, Inc.
Copyright © 2017 by Daniel Cartamil, PhD
All rights reserved. First edition 2017

Cover and book design by Michael Schrauzer
Project management by Deborah Young
Printed in Korea

No part of this book may be reproduced in any form without permission from the publisher. Please direct comments and inquiries to:

Sunbelt Publications, Inc.
P.O. Box 191126
San Diego, CA 92159-1126
(619) 258-4911, fax: (619) 258-4916
www.sunbeltbooks.com

20 19 18 17   4 3 2 1

All photographs are by the author unless noted or in public domain.

Library of Congress Cataloging-in-Publication Data

Names: Cartamil, Daniel, author.
Title: Baja's wild side : a photographic journey through Baja
   California's Pacific Coast region / by author Daniel Cartamil.
Description: First edition. | San Diego, CA : Sunbelt Publications, Inc.,
   [2017]
Identifiers: LCCN 2017001152 | ISBN 9781941384329 (softcover : alk. paper)
Subjects: LCSH: Baja California (Mexico : Peninsula)--Pictorial works. |
   Natural history--Mexico--Baja California (Peninsula)--Pictorial works.
Classification: LCC F1246 .C1817 2017 | DDC 972/.2--dc23 LC record available at https://lccn.loc.gov/2017001152

Photographic reproduction of all images of rock art and historical sites in this work has been authorized by the National Institute of Anthropology and History. Secretaria de Cultura, -INAH, -MEX.

All author's royalties from the sale of this book support Baja California conservation research at the Scripps Institution of Oceanography.

ABOUT THE AUTHOR

DANIEL CARTAMIL, PhD, is an expert in shark biology at the Scripps Institution of Oceanography in La Jolla, California. He has studied sharks since 1993, and has also worked for many years as a fisheries biologist in Alaska and California. Most recently, his passion for the natural world inspired him to devote 10 years to photographing Baja California's untamed wilderness, particularly the Pacific Coast Region, where he conducts research to protect migrating sharks from overfishing. Dan's fluency in English and Spanish, as well as the support of locals, academic colleagues, and artisanal fishermen, have provided him unparalleled access to Baja's spectacular landscapes. Dan is also an environmental consultant and leads adventure photography tours in Baja California (bajaswildside.com). He lives, mountain bikes, and plays Afro-Cuban percussion in Encinitas, California.

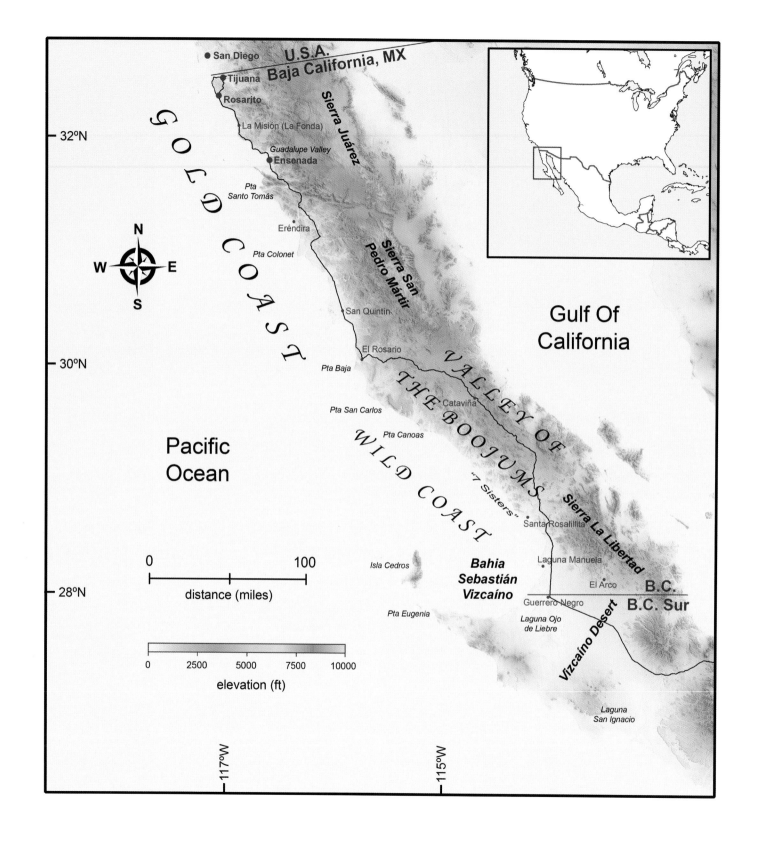

Table of Contents

Foreword by Graham Mackintosh   ix

Preface   xi

Acknowledgments   xiii

Chapter 1: Bahía Sebastián Vizcaíno & Inland   1

    Pacific Coast: Punta Eugenia to Santa Rosalillita

    Vizcaíno Desert

    Sierra La Libertad

Chapter 2: The Wild Coast   27

    Pacific Coast: Santa Rosalillita to El Rosario

Chapter 3: Valley of the Boojums   51

Chapter 4: The Gold Coast & High Sierra   73

    Pacific Coast: El Rosario to Tijuana

    Sierra San Pedro Mártir

    Sierra Juárez

A NOTE ON NAMES & PLACES

There is some confusion about the name "Baja California." Many people use it to refer to the entire Baja peninsula. However, the peninsula comprises two distinct Mexican states, divided by the 28th Parallel (28 degrees north latitude): Baja California (to the north), and Baja California Sur (to the south). For the purposes of this book, "Baja California" refers to the northern state, and "Baja" refers to the entire peninsula.

And what is Baja California's "Pacific Coast Region"? Baja California is "divided" along its east-west axis by a spine of mountain ranges, or sierras. To the east of these ranges, the landscape plummets towards the warm and (usually) calm Gulf of California. To the west, however, the topography slopes more gradually towards the colder, more tumultuous Pacific Ocean. I have arbitrarily designated this western half of the state as the "Pacific Coast Region," an area that includes not only the coastline, but also various inland deserts and mountain ranges.

Finally, the images in this book are arranged into four chapters, each corresponding to a specific subsection of the "Pacific Coast Region." These subsections are also arbitrarily named — you won't find the "Gold Coast," "Wild Coast," or "Valley of the Boojums" on any technical map. This was done for ease of organization and thematic consistency. The Table of Contents and first page of each chapter will tell you exactly what you're looking at, and the map on the left side of this page will help you get oriented.

Now, let's get on with the journey, and explore Baja's "Wild Side"….

BAJA'S WILD SIDE

Foreword

BY GRAHAM MACKINTOSH

Dan Cartamil knows sharks, the people who catch them, and the state of the shark fisheries in Baja California. He holds a PhD from the Scripps Institution of Oceanography, La Jolla, California, and has authored or coauthored over 20 scientific journal articles on the subject. His research projects have led him to spend considerable time on the Baja peninsula and have been instrumental in shaping recent shark conservation measures by the Mexican government.

Beyond his scientific pursuits, his passion and determination to protect and conserve the land he has come to treasure has led to this beautiful photographic journey: *Baja's Wild Side*.

I have been traveling in Baja since 1979 and have written four books about it. Over a span of almost two years (1983–85), I walked much of its coastline and spent many days and nights in the company of fishermen in remote camps. I also spent hundreds of days severely alone traversing an untrammeled wilderness that at times seemed neverending. Desert, mountain, and coast stood assured, eternal.

Graham MacKintosh

Trip over and returning to my native England, I received the "Adventurous Traveler of the Year Award" for the journey, which was recounted in my first book: *Into a Desert Place*.

Having lived so intimately with the beauty of the peninsula and those who have recently impacted it, it was easy to feel sympathy for both. People have lived along and exploited the Wild Coast and its hinterland for thousands of years, but in a way tolerably insignificant, if not harmonious.

In the 1980s, exploring an abandoned mine and railroad half way up a steep mountain of volcanic rock led me to write: "It was easy to summon up the ghosts of the mine workers and imagine the scene at the end of the 19th century. I had to admire the engineering genius behind the railroad. It wasn't a slap that the hand of man had laid across the face of nature, rather a solid monument to human endeavor." [*Into a Desert Place*]

The desert and the sea always seemed capable of reclaiming human inroads and striking them from view.

Having carried just a point-and-shoot 35mm film camera for much of my travels, especially along the stretch of Pacific Coast covered by this book, I was rarely able to take more than a few photographs a day. Sometimes I had no film. Much of what I saw and revered had to be committed to memory — images of colorful coastal grandeur and magnificent desert vistas, fascinating mysterious rock art, missions

BAJA'S WILD SIDE

steeped in history … and everyday life in remote ranches and fishing camps with some of the kindest, most hospitable people on the planet.

Daniel's book with its superb photography brought it alive once more and helped me understand why I was so drawn to the peninsula and its people and compelled against all good sense to walk its dramatic coastline. Indeed, Daniel's keen perspective enabled me to see more than when struggling alone, eyes stinging with sweat and sunscreen, focused a few steps ahead for rattlesnakes and piercing spines.

This timely and important book, while celebrating an astounding wilderness, looks back to a relatively benign and sustainable human impact, and forward to new more existential threats.

Increased tourism, development, and over-exploitation have changed the picture. Whatever balance and harmony there was is ever more threatened by industry, agriculture, sprawling development, golf courses, new roads, and vast swathes of former stunning beauty are now lost beneath ugly, wind-blown trash dumps.

The author's skillful photography captures a moment when the remnants of Baja's Wild Side might still be protected to inspire and uplift the human spirit, or equally may be lost forever.

Graham Mackintosh's Baja Books
Into a Desert Place
Journey With a Baja Burro
Nearer My Dog to Thee
Marooned With Very Little Beer

Preface

In a sense, this book began back in 2006. At the time I was working on my PhD at the Scripps Institution of Oceanography, studying thresher sharks and their migrations along the California Coast. The daily routine went something like this: my team and I would go out in a 16-foot skiff and troll lures back and forth offshore of the coast, sometimes for days, until we managed to hook and reel in one of these 300-pound monsters. After wrestling it to the side of the boat, we would attach an electronic tag to the shark's dorsal fin, and then release it unharmed. Now the hard part: using radar-like tracking equipment, we could home in on the tag's signal and follow the shark around the ocean, documenting its movements and habitat preferences. In our cramped little vessel, through cold, foggy nights and blisteringly hot days, we'd try to follow a shark for 72 hours, which was the approximate limit of our human endurance (not to mention our supplies of food, water, and fuel). Then we'd go home, spend a day or two recovering, and do it all over again.

One of the significant findings of this research was that thresher sharks were migrating across the US–Mexican border into Baja California waters. As is often the case in science, this led to a new series of questions. How far south did thresher sharks migrate, and was there a seasonality to their movements? And from a conservation perspective, were they being fished commercially in Mexican waters, and how did this affect the sustainability of the overall population? These were big questions that I clearly couldn't answer by myself.

I was fortunate at Scripps to have a great mentor in Dr. Jeff Graham, and Jeff introduced me to one of Mexico's preeminent shark experts, Dr. Oscar Sosa-Nishizaki. We agreed to collaborate on a new research project, one that would focus on investigating the catch of thresher sharks in Baja California's coastal shark fisheries. After all, what easier way is there to investigate the geographic distribution of a fish than to look in the nets of local fishermen?

The final member of the team was Omar Santana-Morales, one of Oscar's graduate students. Omar would accompany me on several preliminary scouting missions while collecting data for his own master's thesis. Our plan was to traverse the Pacific Coast in a 4×4, visiting as many fishing camps as possible and interviewing fishermen about their operations and the kinds of sharks they were catching.

From the beginning, it was a strange adventure. Consider this brief journal excerpt from September 12, 2006, my *first day* in Baja. We had departed early in the morning from Ensenada.

"After a long day of driving, it was time to get off the road and find a spot to spend the night. We consulted the map and decided to make camp near Laguna Manuela, where we'd heard reports of a fishing camp that targets sharks and stingrays. So we veered off the Transpeninsular Highway at the little pueblo *of Villa Jesús María, and headed west toward the coast on a badly rutted dirt road that became almost impassable after only a few minutes. Now what? As we got out to assess the situation, Omar noticed there that were strange gray objects all over the desert floor. At first, we really didn't*

know what we were looking at, and it took a few minutes to register....

They were shark skulls. Thousands of them. Tens of thousands. Heaped in piles and strewn about in the dunes as far as we could see, serrated teeth still protruding from their jaws. We'd stumbled onto a shark graveyard of apocalyptic proportions...."

As it turned out, Omar and I would end up taking nearly a hundred expeditions together over the next decade. Laguna Manuela became our main research site, and with Oscar and Jeff as coauthors, we would eventually publish a series of scientific articles based on our studies there.

At the same time, we continued our systematic efforts to survey every fishing camp on the Pacific Coast. This project required a *lot* of backcountry travel, and gave me an opportunity to encounter some of the peninsula's most spectacularly remote terrain. As a life-long photographer, it was only natural that I began to document the entire experience with my camera. And with repeated visits, the beaches, arroyos, mesas, and mountain ranges began to reveal their secrets — their hidden oases, ancient cultural artifacts, and elusive wildlife.

The idea for a conservation photography book was born from the realization that this fragile wilderness is imperiled. It's easy to see Baja as vast, inhospitable, and unchanging; but change is inevitable. Three hundred years ago, Baja's human population numbered in the thousands, consisting of nomadic Indian tribes like the Cochimí. Today, four million people inhabit the peninsula; most of this growth has occurred in the past 50 years. Humanity is expanding into the wild places, and nowhere more so than in the Pacific Coast Region.

So let's take a journey to "Baja's Wild Side," you and I, while we can. These are the images I captured during ten years of research and travel in Baja California. It's my hope that sharing them will do more for the cause of public awareness and conservation than any statistics I might publish in an academic report.

Are you ready? I want to show you what's hidden in the hills.

Acknowledgments

The foremost acknowledgment goes to my research collaborator and good friend Omar Santana-Morales, who has been with me on almost every Baja expedition over the past decade. Omar is the kind of guy you want with you in the backcountry — he is incredibly competent, knows someone in every remote corner of the peninsula, and is a great cook! His passion for exploring and conserving Baja California's wild places was critical to the success of this project. Omar is a researcher and the Director of Marine Projects with the nonprofit conservation group ECOCIMATI (www.ecocimati.org).

Funding for the shark research programs that inspired this photography project includes support from the California Sea Grant and the Save Our Seas Foundation. Special thanks to the Moore Family Foundation which has been supporting the binational shark research program at Scripps Institution of Oceanography for nearly a decade.

Within the academic world, I would like to thank Phil Hastings, Jeffrey B. Graham, Jessie Brooks, and the administrative staff (Scripps Institution of Oceanography); Dovi Kacev (Southwest Fisheries Science Center); Oscar Sosa-Nishizaki and Miguel Ángel Escobeda Olvera (Centro de Investigación Científica y de Educación Superior de Ensenada, CICESE).

Thanks to authors Graham Mackintosh and Harry Crosby for their insights and encouragement, and to Graham for writing the foreword to this book. Thanks to brothers Ian and Nathan Velasquez for guide services, to Espacios Naturales y Desarrollo Sustentable, A.C., (ENDESU) for Peninsular Pronghorn Antelope information, and to Zachary Plopper of WILDCOAST for Baja conservation information. Alejandro Hinojosa kindly provided topographic map files. Thanks to writing coach Nikki Pugh. And heartfelt gratitude to all of the fishermen, rancheros, and other Baja Californianos who have helped us over the years — I have been moved by their generosity and kindness.

On a more personal note, thanks to my family, including my "photography godfather" Peter Arnold, for endless support. Special thanks to my father, Jorge — the outdoor skills he taught me on our countless adventures together were my life-long training camp for Baja. Finally, thanks to Christine Stevens for valuable feedback, support, and marketing.

This book is dedicated to Dr. Jeffrey B. Graham and Miguel "Micks" Olvera — two friends who greatly influenced the course of this project in different ways. A world-renowned pioneer in the field of fish physiology, Dr. Jeff Graham was my PhD advisor at Scripps Institution of Oceanography. While it is well known that Jeff was a brilliant scientist, he was also a wonderful artist and a man deeply concerned with the conservation of marine resources. Micks Olvera was a student when I first met him, but he soon became a valued collaborator on several shark research projects in both the US and Mexico. Micks was known by fishermen throughout Baja California as "the rasta biologist," and memories of his laughter around the campfire make me smile to this day. Both men passed away during the course of this project, but their spirit lives on, subtly etched into every image.

Omar Santana-Morales

Chapter 1 — Bahía Sebastián Vizcaíno & Inland

Named after the Spanish explorer who traversed its waters in the 1600s, Bahía Sebastián Vizcaíno is Baja's largest bay, spanning about 125 miles of shoreline along the Pacific Coast. Its shallow waters extend up to 90 miles offshore, forming an enormous habitat for schools of sharks, rays, and other sea life. This in turn supports highly productive fisheries that are a mainstay of the local economy. Much of the shoreline consists of a convoluted maze of estuaries and wetlands. These provide critical bird habitat and a protective nursery for the young of many fish species. And of course, the region's large inland lagoon — Laguna Ojo de Liebre — is famous for its seasonal habitation by migrating Gray Whales.

Immediately inland, one encounters the northern reach of the Vizcaíno Desert, an arid expanse of flatness stretching south into Baja California Sur. Travelers who are just passing through on the Transpeninsular Highway often find this to be the most uninteresting section of the peninsula. But for those willing to venture off the beaten path, the Vizcaíno Desert reveals its sparse beauty in the form of forests of lichen-draped yucca trees.

To the northeast, the desert gently slopes uphill until it merges with the mountainous interior of the peninsula: the forbidding and nearly impenetrable Sierra La Libertad. Spectacularly scenic, the Sierra and its surrounding foothills are also a trove of anthropological treasures that span the entire range of human history on the peninsula, including ancient rock art of indigenous Indians, the remains of Spanish colonial missions, and more recent mining ghost towns.

Left: A typically foggy morning in the Vizcaíno Desert.

Next Page: Panoramic view of the Laguna Manuela estuary. This unique convergence of desert flora and coastal wetlands gives the shoreline of Bahía Sebastián Vizcaíno its distinctive character.

Above: A religious shrine offers solace and protection to Laguna Manuela's fishermen, who often spend several days offshore in rickety boats fishing for sharks.

Right: A shark fisherman at the Laguna Manuela fishing camp.

Far Right: The foothills of the Sierra La Libertad loom above Laguna Manuela estuary, where a lone fishing boat heads out at first light.

BAJA'S WILD SIDE

Above: Once hunted almost to extinction, Gray Whales have made a tremendous comeback along the Pacific Coast of North America. This colossal skeleton on the shore at Laguna San Ignacio serves as a grim reminder of mankind's past relationship with the whales.

Right: Ospreys (right) are one of the most commonly seen avian predators of Bahía Sebastián Vizcaíno; this one prepares to feast upon a freshly caught juvenile California halibut.

Next Page: Whale-watching draws ecotourists from around the world to Laguna Ojo de Liebre (also known as Scammon's Lagoon), one of the three major lagoons on Baja's Pacific Coast that are seasonal breeding and calving grounds for Gray Whales.

Above: By the 1990s, only about 200 *berrendos* (Antelope = Peninsular Pronghorn Antelope) remained in the wild, primarily in the Vizcaíno Desert. Since that time, efforts to raise berrendos in captivity have been difficult but successful; the current population consists of over 600 animals, most of them located in enormous enclosed ranches outside Guerrero Negro.

Right: Forests of Baja California Tree Yuccas are one of the defining natural features of the Vizcaíno Desert. Domesticated cows, however, are a relatively new invasive species, now found in virtually every conceivable habitat on the peninsula. They can be a major ecological nuisance—displacing native herbivores (including berrendos), destroying vegetation, and trampling sensitive habitat.

CHAPTER 1: BAHÍA SEBASTIÁN VIZCAÍNO & INLAND

CHAPTER 1: BAHÍA SEBASTIÁN VIZCAÍNO & INLAND

Left: Great Mural-style cave paintings that showcase huge humanoid figures on the roofs of desert caves are typical of the indigenous rock art of Baja California Sur. However, as Harry Crosby documented during his now-famous expeditions of the 1970s, they can also be found in the southern reaches of Baja California's Pacific Coast Region, including this dramatic site in the foothills of the Sierra La Libertad. Some Great Murals have been estimated to be up to several thousand years in age; they are powerful reminders of humanity's primitive past in a setting of vast unspoiled wilderness.

Next Page: A more straightforward view of the mural allows identification of several animal forms, including fishes, a Bighorn Sheep, and a berrendo. Colorful and convoluted patterns of weathered granite add to the beauty of the paintings. To the right of the page, views from the inside and exterior of the cave.

BAJA'S WILD SIDE

One of Baja's many ghost towns, Pozo Alemán offers a glimpse into the Vizcaíno Desert's recent past. This now-abandoned gold mining site had its heyday around the year 1900. Note the crumbling caves dug under the main structure (as well as various locations around the mine); they apparently provided cool, if not particularly safe, underground dwellings for mine workers.

CHAPTER 1: BAHÍA SEBASTIÁN VIZCAÍNO & INLAND

The ravages of time slowly disintegrate this adobe building at Pozo Alemán.

BAJA'S WILD SIDE

Above: Misión Santa Gertrudis, founded by Jesuits in 1751, occupies a remote and mountainous corner of the Vizcaíno Desert.

Left: A rare Christian-themed petroglyph of the Holy Cross etched into a boulder near Misión Santa Gertrudis. Was it carved by a Jesuit padre or by one of the indigenous Indians who painted the mysterious figures (inset) found in a nearby cave overlooking the mission? The figures, several meters in height, are painted half black and half red; some have speculated that this was meant to symbolize the duality of human nature.

Right: A ranchero and his guitar, Misión Santa Gertrudis.

CHAPTER 1: BAHÍA SEBASTIÁN VIZCAÍNO & INLAND

Above: The mission bells at Misión Santa Gertrudis.

Left: A lonely cemetery near Misión Santa Gertrudis.

A colorful series of rock art paintings is scattered over several hundred feet of cliff face at a site towards the northern end of the Sierra La Libertad. Of particular significance is the humanoid figure, or *mono*, seen on the next page; the right half of the figure, painted black, is badly faded. This is the northernmost *mono* that has been found, marking the known geographical extent of the Great Mural-style paintings. As will be seen in the following chapters, the rock art of northern Baja California is generally dominated by more abstract geometric figures, such as the line drawings on the top of this page. There are, of course, numerous exceptions like the vibrant red and yellow flower seen to the right.

Left: An unpaved road leads through wilderness in the foothills of the Sierra La Libertad.

Right: Spiral architecture at Misión San Borja's stone church, completed in 1801.

Below: Partially preserved remains of the original adobe mission at San Borja.

Chapter 2 The Wild Coast

Just above Bahía Sebastián Vizcaíno, there is a pristine stretch of coastline almost 140 miles long; I call it the Wild Coast. Wild, because it is sparsely inhabited by humans. Wild, because it is a rare, undisturbed sanctuary for coastal wildlife. Pelicans, seagulls, and terns patrol the shore, and the bays teem with lobster, halibut, and clams. Coyotes and vultures roam the high tide line, scavenging upon the carcasses of stranded marine mammals.

Most of all, this coast is wild because of its awe-inspiring and diverse topography. At its northern end (just below Punta Baja), the Wild Coast forms a wall of imposing bluffs: sheer cliffs of soft, eroding sandstone that occasionally collapse onto the cobblestone beaches below. At Punta San Carlos, spectacular volcanic mesas begin to dominate, plunging into the sea from heights of over 1,500 feet. Then, at Punta Canoas, the coastline transitions to lovely shell-paved beaches and dune fields, punctuated by occasional rocky headlands. This idyllic seascape extends south, through the Seven Sisters region, all the way to the fishing village of Santa Rosalillita.

Technically, the Wild Coast is the western edge of the Valley of the Boojums (the subject of Chapter 3). When combined, these two subsections of the Pacific Coast Region extend over a third of the state's land; this land is designated as a Flora and Fauna Conservation Area, making it one of Mexico's largest protected zones. Nevertheless, it is not always clear what is protected, and to what extent. According to WILDCOAST, an international organization that is actively conserving the area: "Despite its remoteness, parcel sales and subdivisions, tourism, mining, and industrial development are continuous threats to the region's ecological integrity."

Next Page: A landlocked fleet of dilapidated wooden *pangas* (the Mexican term for small traditional fishing vessels) adorns the entrance to the tiny fishing village of San Carlos.

Left: If any avian species is emblematic of Baja's Wild Coast, it must be the Brown Pelican. Whether diving on a school of anchovy from the heights, or skimming above the waves toward some unknown destination, these predators exude grace and power.

CHAPTER 2 THE WILD COAST

Above: A classic Baja signpost marks the San Carlos "exit" off the Transpeninsular Highway.

Left: Low-hanging fog adds an almost mystical quality to this early morning scene along the coastline at Punta San Carlos.

CHAPTER 2 THE WILD COAST

Left: Along the edge of a boulder-strewn volcanic mesa, Omar and I encountered this shell midden—a scattered pile of abalone and clam shell fragments, bleached white and weathered from perhaps several hundred years of exposure to the elements. A second midden can be seen in the background. Presumably this was a gathering or ceremonial site for the indigenous people of the region. But why make the arduous trek to the top of this mesa with their shellfish harvest? The cliff walls above the midden abound with petroglyphs, such as the checkerboard patterns etched into the boulder (just to the left of the cactus) in this photo.

Below: This petroglyph, found just above the shell midden, may offer a clue to the mystery. The indigenous tribes of Baja California were decimated by contact with the Spanish, who came by sea seeking riches and religious converts. An interesting hypothesis to consider is that this petroglyph represents the aft end of a Spanish galleon departing upon the waves, which would have been a powerful symbol to a beleaguered native population. Perhaps this vantage point high above the sea was advantageous in monitoring the activities of the Spanish fleet. We may never know....

CHAPTER 2   THE WILD COAST

Above: Various coastal critters leave their distinctive "foot"-prints along the shoreline.

Left and Following Page: The (approximately) 50-mile stretch of beaches, rocky points, and occasional fishing camps known informally as Seven Sisters is one of the most pristine coastlines on the Baja peninsula.

CHAPTER 2 THE WILD COAST

Left: Long after the sun has set, a sliver of crescent moon descends toward the horizon just south of Puerto Catarina.

BAJA'S WILD SIDE

Above: An otherworldly scene is created when the orange-tinged cliffs of Punta Canoas reflect the last rays of the setting sun.

Right: While camping on the beach near Punta Canoas (towards the left side you can see the truck, campfire, and tents), we were treated to an epic evening of lightning strikes in the distance. Unfortunately, they didn't stay in the distance, and by 10 pm lightning was falling all around us, and our tents were completely flooded. There was nothing left to do but get into the truck, open a bottle of wine, and hope for the best....

Left: Another beautiful morning on the Wild Coast, south of Punta Canoas.

Below: An "adventurer's shrine" of beach stones, rusty tin cans, shells, car parts, bleached bones, and tequila bottles marks a favored camping spot used by an intrepid few who have braved the rutted dirt road along the Wild Coast, south of Punta Canoas.

BAJA'S WILD SIDE

These soft sandstone cliffs near Puerto Catarina crumble to the touch, making photography a risky proposition. The intricate patterns, sculpted by wind and water, represent a different type of Baja "rock art."

BAJA'S WILD SIDE

A coastal arroyo near Punta Baja glows after an early evening shower.

CHAPTER 2 THE WILD COAST

Near-vertical bluffs along the Wild Coast, south of Punta Baja.

CHAPTER 2 THE WILD COAST

A turbulent Pacific Ocean reflects the dawn sky's pastel shades just south of Punta Baja. For those heading south, this is where the Wild Coast begins....

Chapter 3 Valley of the Boojums

The Transpeninsular Highway makes a hairpin turn — a 90-degree inland twist — in the small town of El Rosario. Seasoned Baja travelers understand its significance as a jumping-off point, a radical departure from coastal towns and cool ocean breezes. To many, this is the where the "real" Baja begins. Within less than 30 minutes, you have entered one of the planet's most unusual and surreal landscapes — the Valley of the Boojums.

Rolling coastal hills transform to a seemingly endless succession of boulder fields, mesas, and canyons. Chaparral and agave stalks give way to a dense jumble of cholla, giant Cardón, and Ocotillo. But one particular species dominates, not by size or number, but by the sheer force of its persona: the exquisitely bizarre Boojum Tree. Resembling giant, skinny, upside-down carrots, Boojum Trees may grow straight as an arrow, or contort and twist into an amazing diversity of shapes. Their presence lends an almost mystical quality to the landscape.

The average elevation of the Highway as it cuts through the Valley of the Boojums is about 2,300 feet — this is Baja California's version of high desert. But the valley also encompasses a vast and largely unexplored wilderness that descends westward to the Pacific, penetrated only by a handful of rugged dirt roads. Rock art of ancient indigenous people is scattered throughout the region, which is also home to Misión Santa María, one of Baja's most remote and picturesque adobe missions.

Perhaps I'm biased, but I find that the remoteness, beauty, and crystalline silence of the Valley of the Boojums make it one of the most inspiring places on the peninsula, if not the entire planet. If you don't believe me, turn the page and decide for yourself.

Left: On a moonless night in the Valley of the Boojums, the delicate front arch of Misión Santa María still stands as it has through the centuries, illuminated by my headlamp against a backdrop of infinite stars.

Next Page: Dating back to 1768, Misión Santa María is one of most beautifully situated and remote of Baja's remaining adobe missions. The same water source that supports large palm groves around the mission transforms into a series of oases as it descends westwards towards the Sea of Cortez.

CHAPTER 3 VALLEY OF THE BOOJUMS

Previous Page: Massive Cardón cacti, some growing to 60 feet in height and weighing over 10 tons, dominate the landscape near the tiny desert outpost of Cataviña.

Far Left: Extreme temperature fluctuations in the high desert cause the exterior of granite boulders to flake off in bizarre patterns.

Left and Below: The Valley of the Boojums can be a windy place, but early morning often brings a deep, still silence broken only by birdsong or the chirp of an insect.

CHAPTER 3 VALLEY OF THE BOOJUMS

Above: This mischievous Boojum Tree seems to form an upside-down peace sign. The clusters of yellow flowers at the top of the tree typically bloom in late summer.

Left: Close-up view of a Boojum Tree's trunk, showing the thick green bark and thorny branches it uses to deter insects and herbivores.

Next Page: A celebration of the many forms and moods of the mysterious Boojum Tree.

CHAPTER 3 VALLEY OF THE BOOJUMS

Left: Countless arroyos descend from the Valley of the Boojums to the Pacific Coast, winding through remote and nearly inaccessible wilderness.

Below: The indigenous people of Baja California's Pacific Coast Region have left a rich legacy of rock art, much of it concentrated in the Valley of the Boojums. Rock art sites are often located near water sources; these red and yellow paintings were found above a semi-permanent *pozo* (water hole) in Arroyo La Bocana.

CHAPTER 3 VALLEY OF THE BOOJUMS

Left: As the sun's first rays hit the surrounding peaks, a tenacious grove of half-burned palm trees rustles loudly in the windy canyon below. Angry bees begin to swarm, driven to a frenzy by thirst. Dawn finds me clinging to the side of a cliff with one hand and swatting bees away with the other, pausing only long enough to photograph these petroglyphs, which appear to my imagination as a patch of rattlesnake skin (left) and a centipede (right).

Above: This ring of stones may have served as a gathering place for native Indians. Sharpened flint chips and rock art found within the immediate vicinity of the ring suggest that it was indeed created by an ancient indigenous people.

BAJA'S WILD SIDE

Above: A simple petroglyph consisting of (what appears to be) four points within a sun.

Right: Fascinating examples of petroglyphs that depart from the basic geometric shapes format. The mysterious figure toward the left *could* represent an owl, but how does your imagination interpret the middle two figures? Note how the rock facing towards the right of the picture is studded with hundreds of tiny pits.

CHAPTER 3 VALLEY OF THE BOOJUMS

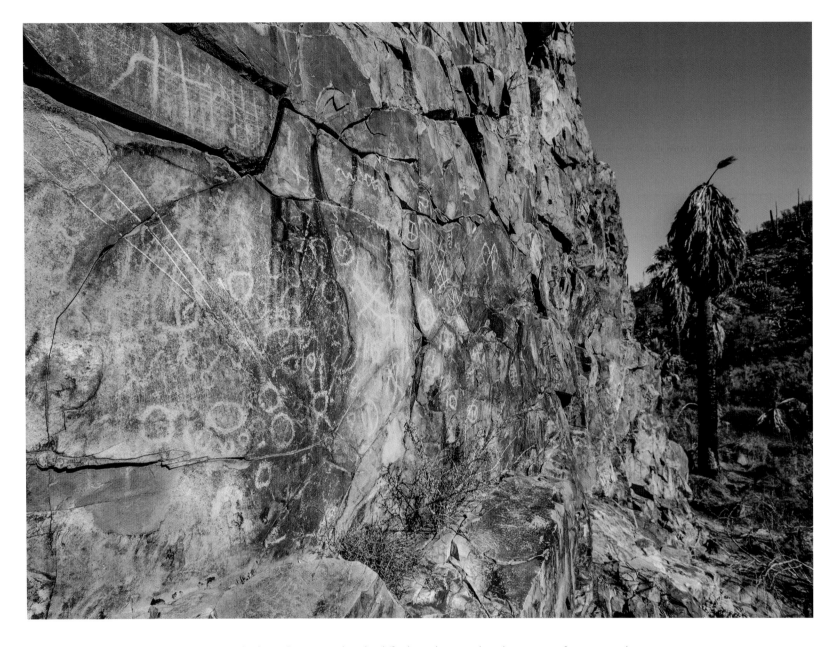

Above: The rock art of Baja California's Pacific Coast Region tends to be difficult to photograph in the context of its surroundings since it is usually located in deep shade regardless of time of day. Perhaps this was a by-product of the artist's desire to reduce sun exposure during what was likely a time-consuming process. This wall is an exception — a mosaic of hundreds of lines and circles exposed to direct sunlight for much of the day. However, they are much more faded than nearby shade-protected petroglyphs, implying sun damage or substantially greater age.

Right: This rock face showcases the classic abstract geometry that characterizes much of northern Baja's rock art.

BAJA'S WILD SIDE

The ruins of Misión San Fernando de Velicatá, established in 1769 and abandoned in 1818, include this well-preserved water storage dam and aqueduct. The decayed adobe mission itself, and a number of interesting petroglyphs that predate the mission by hundreds of years, are close by.

CHAPTER 3 VALLEY OF THE BOOJUMS

Along the length of the Transpeninsular Highway, hundreds of shrines commemorate those who have lost their lives on this treacherous road. *Descanse en paz*, Hector.

Chapter 4   The Gold Coast & High Sierra

Crossing the border from the US into Baja California is a minor geographic shift, just a few feet across an arbitrary political line. Yet the cultural shift is radical and immediately obvious. This is a different world, one of mariachi music, street vendors, sombreros, bright colors, and friendly smiles. Even the air smells different: a combination of sea mist, taco stands, and burning garbage. We have now entered the most populated and commercially exploited portion of the peninsula. Seemingly an odd chapter to close a book that has thus far emphasized remote landscapes. But there is a reason tourists and adventurers from around the world come to this place. For hidden away amongst the trappings of civilization, there are still pristine jewels of natural beauty.

The Gold Coast is a narrow seaside corridor that straddles the Transpeninsular Highway from El Rosario to Tijuana. Much of this passage is nondescript and hilly, covered with shrub-like coastal chaparral and cacti. There are also dramatic stretches of rocky shore, where the pounding surf sends up a perpetual mist that often foiled my photographic efforts. Unfortunately, much of the Gold Coast's natural habitat has been disturbed by agriculture and real estate development. Nevertheless, there are dormant volcanoes, shipwrecks, estuaries, mission ruins, isolated beaches, and even a fascinatingly historical wine country to be explored.

Moving inland, the terrain ascends toward one of the region's great surprises — the High Sierra. After all, snow-capped and heavily forested mountain ranges, dotted with alpine meadows, are not the kind of landscapes typically associated with Baja. Two such ranges tower above the surrounding countryside: Sierra Juárez and Sierra San Pedro Mártir, attaining elevations of over 6,000 and 10,000 feet respectively. To paraphrase author Graham Mackintosh, they are islands in the sky.

CHAPTER 4   THE GOLD COAST & HIGH SIERRA

Right: Looking south over the chasm of Cañon El Diablo, towards the peninsula's highest peak, Picacho El Diablo, which tops out at an impressive height of 10,157 feet. To the west, dense pine forests cover the sierra for hundreds of square miles.

Below: Shrine to San Pedro Mártir, found within the National Park that bears his name.

BAJA'S WILD SIDE

A sweeping, hazy vista of the Sierra San Pedro Mártir foothills, taken at an elevation of approximately 6,500 feet. The white surface in the foreground is rock, not snow, although snow is common in the sierra during winter months.

BAJA'S WILD SIDE

Once populous throughout North America, the California Condor was extinct in the Mexican portion of its range by the 1980s. A binational project, spearheaded by the San Diego Zoo and Mexican government partners, has been working to reintroduce these oddly charismatic birds to the Sierra San Pedro Mártir, where 35 free-flying condors now reside.

CHAPTER 4 THE GOLD COAST & HIGH SIERRA

A California Condor soars high above the arid foothills of the Sierra San Pedro Mártir. Biologists can track the movements of newly released condors through the use of satellite transmitters (visible in this picture as white bands on the wings).

BAJA'S WILD SIDE

Above: The tiny and charming chapel at Rancho Meling, a guest house and working ranch at the base of the Sierra San Pedro Mártir's western flank.

Right: As seen from the relatively low elevation of 2,500 feet, the topography of the Sierra San Pedro Mártir changes rapidly with increasing altitude, forming discrete layers of color, flora, and terrain.

BAJA'S WILD SIDE

Above: Laguna Hanson is a rare jewel—a pine-fringed alpine lake found at an elevation of over 5,000 feet in the Sierra de Juárez mountain range.

Right: The Bobcat is a common but elusive predator of the High Sierra.

CHAPTER 4   THE GOLD COAST & HIGH SIERRA

Left: Perched above the ocean, the quaint hotels and restaurants of La Fonda have been popular with Southern California tourists for decades; they come to enjoy the charm and natural beauty of "Old Mexico," just 40 minutes south of the US border.

Above: Slowly but surely, the prime real estate that is Baja California's Pacific Coast Region undergoes commercial development. Unfortunately, this often leads to pollution, habitat loss, and over-exploitation of marine resources. Practicing sustainable development will be a major challenge for the Gold Coast in the 21st century.

BAJA'S WILD SIDE

Valle de Guadalupe is Baja California's answer to Napa Valley. The highly fertile region abounds with vineyards and historic sites, and has become a major tourist destination for Southern Californians and residents of northern Baja.

CHAPTER 4   THE GOLD COAST & HIGH SIERRA

Rio de Janeiro may have Corcovado, but Baja Californianos have their own "Redeemer Upon the Hill" just south of Rosarito.

87

CHAPTER 4   THE GOLD COAST & HIGH SIERRA

The spiritual life of Baja Californianos is exemplified by their devotion to La Virgen de Guadalupe, who is said to have appeared to the Indian peasant Juan Diego in Mexico City in 1531. Each year in mid-December, several hundred of the faithful undertake a 10-day pilgrimage through the countryside, beginning and ending in Tijuana. Along the way, they are fed and housed by churches and devotees of La Virgen. This image shows the front of the procession. The first car, draped in white, carries an effigy of *El Niño Divino* (Jesus as a divine child), followed by an elaborate mobile shrine to La Virgen, surrounded by cacti and roses. On the day this picture was taken, I walked with the pilgrims for several hours, and they never once stopped singing.

CHAPTER 4   THE GOLD COAST & HIGH SIERRA

Above: Known locally as *El Barco*, this rusty hulk is all that remains of the vessel *Isla del Carmen*, wrecked at Punta San Jacinto in 1981.

Left: Colorful pangas (fishing vessels) float at anchor in the kelp-choked water off Punta Santo Tomás.

CHAPTER 4   THE GOLD COAST & HIGH SIERRA

Above: Sea Lions need a place to hang out and get out of the surf, and it's not hard to tell who the top dog is on this rock outcrop just offshore of the fishing and agricultural village of Eréndira.

Left: Typical coastal flora on a hillside, looking north towards the steep cliffs of Cabo Colonet.

BAJA'S WILD SIDE

Right: At first glance, San Quintín is a sprawling, dusty agricultural town with little obvious appeal. But natural beauty abounds just off the main road. Bahía San Quintín is a lovely bay that supports a thriving aquaculture industry (the black lines in the water are oyster farms), and is surrounded by a series of extinct volcanoes known as the San Quintín Volcanic Field. The cinder cone towards the left of this image is the rather unimaginatively named Volcan Sudoeste or Southeast Volcano. This shot was taken from the flank of a higher cone (Cerro Kenton) during a fantastic late afternoon light show above the Pacific.

Below: Oyster farmers in Bahía San Quintín pass next to a raft of resting cormorants below the yawning crater of Volcan Sudoeste.

Next Page: Inspiring views from the lichen-encrusted rim of Volcan Sudoeste.

CHAPTER 4   THE GOLD COAST & HIGH SIERRA

Wetlands and tidal creeks, Bahía San Quintín.

BAJA'S WILD SIDE

Fishing village, Puerto Santo Tomás.

Epilogue

JOURNAL ENTRY: FEBRUARY, 2016

I am camping at Puerto Santo Tomás, an out-of-the-way little fishing village at the end of a 22-mile dirt road. Not just any dirt road, but the *fun* kind. The kind of rutted, muddy, cliff-clinging dirt road that keeps the casual tourist *out*, and irresistibly lures the Baja explorer *in*. I've had the whole place to myself for two nights, but today three other Americans pulled in. John is a retired lawyer hauling a comically oversized trailer; Rick and Sonya are Pacific Northwest hippies driving all the way to Cabo in an old VW van. Like me, they saw this place on a map and said, "I wonder what we'd find if we went *there*?"

Well, for one thing, we've found each other, and experience has taught me that most of the people you meet in rural Baja are good people. It's a typically blustery Pacific-Coast-in-the-winter evening: damp and cold. And we all meet up behind John's trailer just before sunset to watch the light show and share a bottle of Guadalupe Valley zinfandel. All we need now is a fire, but it's just too windy. Rick and Sonya are newcomers to Baja and they're surprised by the weather. But I tell them that's the Pacific Coast. It's *usually* this way! We sit by the edge of a bluff watching the surf pound into the ominous black rocks below, trading stories by the light of a battery-powered lantern. And as night descends, my new friends are surprised by another discovery. Despite our clear line of sight at least 10 miles down the coast to Punta San José, there isn't a light to be seen, not even in the little fishing village to our left.

We're only 20 miles from Ensenada — one of Baja's largest cities — yet enveloped in darkness, shrouded in secrecy, surrounded by natural beauty. Contemplating the humble plywood shanties inhabited by local fishermen, we talk about how such a place would be covered in multi-million dollar homes in the US. We are overtaken by a feeling of awe, an overwhelming gratitude to be here and to partake of this silence.

What will it look like here in 100 years? For that matter, if the road in is paved, what will it look like in just 20 years? The future is so uncertain, and it's easy to be pessimistic. If we don't preserve these wild places, we risk much more than a polluted environment and a loss of pristine habitat. We risk our very humanity. For we know there is an inherent and incalculable value to unspoiled nature. To spaciousness and tranquility. To a night sky so full of stars that the darkness beyond is barely visible. To Baja....